W9-AUA-891

DISCARD
PORTER COUNTY
LIBRARY SYSTEM

P J '89 | | 575

J
669 Fod $13.95 BTSB
Fodor, R. V.
Gold, copper, iron

Portage Public Library
2670 Lois Street
Portage, IN 46368

OCT 1 9 1989

GOLD,

COPPER,

IRON

How Metals Are Formed, Found, and Used

R.V. Fodor

— an Earth Resources book —

ENSLOW PUBLISHERS, INC.

Bloy St. & Ramsey Ave.	P.O. Box 38
Box 777	Aldershot
Hillside, N.J. 07205	Hants GU12 6BP
U.S.A.	U.K.

PORTER COUNTY PUBLIC LIBRARY SYSTEM

P J '89 11575

Copyright © 1989 by R. V. Fodor

All rights reserved.

No part of this book may be reproduced by any means without the written permission of the publisher.

Library of Congress Cataloging-in-Publication Data

Fodor, R. V.
  Gold, copper, iron.

  (An Earth resources book)
  Bibliography: p.
  Includes index.
  Summary: Discusses the geologic formation of metal ores, the types of tools scientists use to find such deposits, and the many ways in which they are used.
  1. Ore deposits—Juvenile literature. [1. Ore deposits. 2. Mines and mineral resources] I. Title. II. Series.
  QE390.F63   1989   669   87-24464
  ISBN 0-89490-138-9

Printed in the United States of America
10 9 8 7 6 5 4 3 2 1

ILLUSTRATION CREDITS

AMAX, pp. 15,64,72,73,74; ASARCO. Inc., p. 62; Atlas Copco Canada, Inc., p. 61; Barringer Research. pp. 56,57; Becor-Western, p.63; Ted Bornhorst, p. 60; Cleveland-Cliffs Iron Co., pp. 73,74; COMEX, p. 69; J.D. Griffs, U.S. Geological Survey, p. 20; John Holden, pp. 8,9,23,26,27,29,30, 32,39,40,41,43,44,46,48,64; Homestake Mining Company, pp. 7,76,83; Courtesy of INCO Limited, pp. 49,75,77; Institute of Scrap Iron and Steel, Washington, D.C., p. 85; Kennecott Copper Corporation Photograph, by Don Green, p. 82; Kerr-McGee Corp., pp. 54,59,61; McPhar, Ontario, p. 57; National Aeronautics and Space Administration, pp.13, 54,86; National Air Photo Library, Canada, p.6; North Carolina State Archives, p. 12; North Carolina State Archives (originally published in *Harper's Magazine*), pp. 33,65; Troy Pewe, #2348, 1966, p. 53; F.L. Quivik, Butte, Montana, p. 80; Reynolds Aluminum, pp. 16,85; Scintrex; Ontario, pp. 53,55,56; U.S. Geological Survey, pp. 22,38,53,58; Woods Hole Oceanographic Institution, pp. 31,36,37,42: World Museum of Mining, Butte, p. 24.

Contents

1

What's in a Rock?

Canadian prospector Jule Cross lived the winter of 1931 on frozen Steep Rock Lake in the Ontario wilderness. He spent long, cold days operating machinery to drill through the ice and into mud at the lake bottom. His goal was to find the treasure he believed to be buried below: enough iron in the rocks to make a billion cars.

For weeks Cross fought blizzards and frigid temperatures that dipped to −45°C (−50°F). At one point his ears froze. But his determination paid off. By drilling into the rock underlying the mud, Jule Cross discovered that Steep Rock Lake hid one of the world's richest accumulations of iron ore.

Unfortunately, the prize was impossible to reach. The iron-rich rock was buried not only by a deep lake, but also by 400 feet (125 m) of lake-bottom mud. There was little Jule Cross could do at that time to recover, or mine, the iron.

Yet acquiring that iron was to become essential for both Canada and the United States. During World War II, ten years after Cross's discovery, enemy ships were sinking iron-ore shipments coming to North America from other countries.

Both Canada and the United States therefore needed new sources of iron. With supplies running short, the two governments had little choice but to attempt extracting the ore beneath Steep Rock Lake—no matter what the cost.

Engineers executed the first step in the monumental task by rerouting the Seine River, which flowed into Steep Rock Lake. Second, they constructed channels and dams and excavated tunnels to drain part of the lake. Finally, they scraped away the thick mud layer that had been the bottom of Steep Rock Lake.

The work to reach the iron-rock was nearly equal to building another Panama Canal. The job took several years and cost over 50 million dollars. But the payoff continues. Steep Rock Lake remains the most important deposit of iron in Ontario since mining began at the lake in 1945.

An aerial view of Steep Rock Lake iron-mining district, northern Ontario, from several miles directly above.

Man has always gone to great lengths to acquire metal from the earth. He has searched, and still searches, for rock rich in iron, for example, to use for building bridges and box cars. He looks for gold and silver to use in jewelry and electronics. He sometimes searches for copper and aluminum, or for metals we call strategic because of their importance in national defense.

Geologists, scientists who study the earth, have technical descriptions for the rocks that provide metals. They recognize that all matter, including rock, is made of atoms of elements such as oxygen, carbon, iron, aluminum, and calcium. Atoms

Mining for gold in northern California.

of one element generally occur in combination with those of other elements. For example, atoms of tin may be chemically combined with oxygen as tin oxide. And lead may be bound to sulfur as lead sulfide. Some metal atoms, however, occur in pure form without other elements. Examples are native copper and native gold.

Atoms that are packed together may make up a mineral. To actually be a mineral, the matter cannot be part of animal or plant life. This defines minerals as inorganic earth material, and it means that oil and coal are not minerals. They are

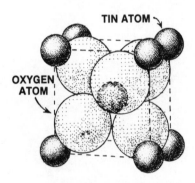

TIN ATOM

OXYGEN ATOM

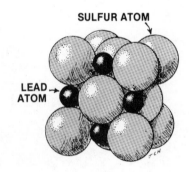

SULFUR ATOM

LEAD ATOM

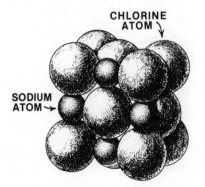

CHLORINE ATOM

SODIUM ATOM

Various combinations of certain atoms, such as tin and oxygen, make up earth material, or minerals, that provide metals. Table salt, or halite, is a mineral made of an orderly arrangement of sodium and chlorine atoms.

8

organic substances, formed from the remains of ancient animals and plants. On the other hand, in some business and legal matters, oil and gas may be considered "mineral resources" even though they are not actually minerals.

To be a mineral, the substance must occur naturally and have a definite chemical composition. Put another way, the atoms of the various elements in minerals are in specific proportions. There may be one atom of iron for every two sulfur atoms, for example. Besides providing particular chemical compositions, these orderly arrangements of atoms give minerals definite internal structures.

The common mineral halite is a good illustration of the chemical and physical properties of minerals. Halite, better known as table salt, consists of the elements sodium and chlorine, and the atoms of each are arranged at the corners of cubes. These cubes represent halite's internal structure: one sodium atom alternates with one chlorine atom. The cube shapes of salt grains are visible with the aid of a magnifying lens.

Mineral grains or crystals make up rocks, just as stacked bricks form brick walls. Some minerals in rocks are large and easy to see. Others are microscopic. A large amount of a val-

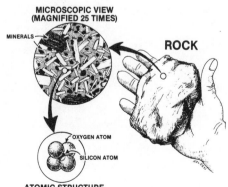

Rocks consist of minerals—and minerals, in turn, are made of various atoms in orderly arrangements.

9

A photograph of a volcanic rock as seen through a microscope. Each of the mineral grains is about 0.039 inch (1 mm) in size.

uable mineral that occurs in or at the surface of the earth is called a mineral deposit. Where a mineral deposit can be mined for financial gain, geologists refer to it as an ore deposit. The rock containing the valuable mineral is ore.

Minerals have been important to man for thousands of years. The types most necessary, however, have changed, depending on the needs and skills of the different societies throughout time. For example, early people used clay for making pottery. Some primitive groups used lead and copper minerals to obtain pigments for paints, rouge, and even eyeshadow. They formed weapons from hard minerals such as quartz.

Extracting metals from rock began about 5000 B.C., when Mediterranean cultures discovered that lead, tin, and copper made better tools and weapons than stone did. During this Copper Age, as it is called, people melted these metals from the ores by using hot charcoal in crude earthen ovens. Afterwards, they shaped the hot metal into axes or swords.

Some cultures learned to mix copper with tin to form an alloy, or metal mixture, called bronze. A small amount of molten tin combined with copper made an alloy that was harder than copper alone.

Historians believe that progression from the Stone Age into the Bronze Age affected how civilization evolved. For instance, armies equipped with bronze swords and armor conquered those that had equipment made only of copper.

Gold and silver also have a long history of use. As far back as 4000 B.C. the Egyptians formed gold into ornaments to adorn the tombs of their dead. They removed gold from rock by crushing and grinding the rock to a fine powder. Then they washed the powder to separate the heavy gold dust from the lighter rock dust. Skilled metal workers could flatten gold into leaves so thin that it took hundreds of gold leaves to make a half-inch (1-cm)-thick pile.

Early civilization advanced in using metallic minerals when it learned to replace bronze weapons with those of iron. The Iron Age was born about 1200 B.C. It put metal into the hands of practically everyone because iron-bearing minerals were relatively common across the civilized world—Europe, northern Africa, and the Middle East.

The metals used thousands of years ago are still important. Copper, for instance, is a major ingredient in electrical wiring and coins. Tin is used in food containers, plumbing, and sheet metal. Gold and silver jewelry is as popular today

Mining gold at Gold Hill, North Carolina, in the late 1800s.

as ever. But the space and computer industries call for special metals such as titanium, manganese, zirconium, and germanium. They are used to make lighter and stronger metals for aircraft, and for sophisticated electronic parts, like computer chips. As technology advances, the needs increase for special metals and alloys made from them and for the ability to locate and mine the metals.

The 1500s heralded mineralogy and economic geology, the sciences of minerals and mineral deposits. Both beginnings are owed to books written by the German physician and

The space shuttle program uses a variety of metals in the body of the shuttle and in its electronic components.

mining geologist Georgius Agricola. Inspired by mining activity around him in Saxony (Germany), Agricola composed ten books on the subjects of geology and mining. His *De Re Metallica* (*On the Subject of Metals*) became the most famous. Agricola's explanations of how metals formed from hot solutions inside the earth are the foundation for today's theories about ore formation. *De Re Metallica* stood as the most important guide to miners for two hundred years following its publication and earned Agricola the title "father of mineralogy."

After Agricola's time, the most important event in mineralogy was the discovery of X rays. It was 1895 when W.C. Roentgen first observed energy rays that were able to pass through solid objects. Not knowing at the time what kind of invisible rays he had found, he simply called them "X." But the name made little difference. Within a few years, scientists had harnessed the mysterious rays to help them determine the atomic arrangements of minerals. That is, X rays enabled them to "see" the atomic structures of minerals.

Both X-ray analyses and special laboratory methods for determining chemical compositions are widely used in the science of mineralogy. To date, they have helped scientists identify over three thousand different mineral types. And several dozen new minerals are discovered each year. However, only about fifty of the thousands known are truly important to our daily lives.

People rarely see the useful minerals, but they are nevertheless present in a variety of forms that man has changed them into. Supermarkets sell food in metal cans made from iron and tin, and we pay for the food with coins containing copper, zinc, and nickel. A person preparing dinner may cook with aluminum pots. Airplanes made of various metals, such as titanium and molybdenum, transport many people each day.

Special instruments control the application of x-rays to provide information on chemical compositions and atomic structures of minerals.

Rolling out sheets of molybdenum-bearing metal for application in modern machinery and technology.

Bricks, or ingots, of aluminum before being made into beverage cans, aircraft, and automobile parts.

All the different metals important to living comfortably come from various places within the earth where they are concentrated, or present in more than normal amounts. How did these metal deposits form? How are they found? And how are people going to protect these deposits, find more, and use them in wise and conservative ways?

2

The Birth of Metallic Minerals

Mauna Loa is the highest mountain in the world. Forming the center of the island of Hawaii, Mauna Loa is a volcano stretching 6 miles (10 km) above the Pacific Ocean floor. It would take twenty-one Empire State buildings stacked upon one another to match that height.

Although it is a volcano, Mauna Loa seldom erupts. The intervals between eruptions of molten rock, or magma, can be several years. But during periods of rest, Mauna Loa gives signs that the magma inside is slowly pushing up toward the volcano summit.

In 1980, for instance, five years after its previous eruption, rising magma began to inflate Mauna Loa. It takes special tiltmeters to detect the swelling of volcanoes, but the behavior is like that of air slowly inflating a balloon. Numerous earthquakes accompanied the movement of magma inside. They were small tremors except for one occurring beneath Mauna Loa's southeast flank on November 16, 1983. It rocked the city of Hilo 30 miles (48 km) away. Around that time, ground temperatures measured along a fracture on Mauna Loa's

19

slope periodically increased to about 30°F above normal, and gas measurements showed greater than normal amounts of hydrogen escaping from the volcano.

Fireworks began on March 25, 1984, when red-orange fountains of lava—which is what magma is called when it reaches the surface—shot from fractures near the summit. Within hours, new fractures opened on Mauna Loa's northeast flank and lava spewed from those. The lava began a journey toward Hilo, moving like honey sliding down a gentle slope.

After one week, the lava stream had traveled over 5 miles (8 km). By the second week, concern developed that it would

Lava, or molten rock material, flowing down the slope (from top of photo to bottom) of Mauna Loa volcano, Hawaii, in April 1984.

enter Hilo and destroy part of the city. In the third week of continuous eruption, Hilo residents in the path of the advancing lava lived in fear of losing their homes.

Then, when lava was only 5 miles (8 km) from the city limits, Mauna Loa's fire burned out. Twenty days after its start, only a long trail of steaming volcanic rock remained as evidence of the fiery event.

Volcanic eruptions like this one dramatically show that molten rock rises from depths to form new rock. Mauna Loa, for example, has piled layer upon layer of magma during the past million years to build to its present height of over 2 miles (3 km) above sea level. Its magmas contain chiefly the ele-

A plume of volcanic gas rises from the flank of Mauna Loa volcano, Hawaii, in April 1984.

ments silicon, aluminum, oxygen, magnesium, iron, and calcium, and they have cooled to rock made largely of the nonmetallic minerals olivine, pyroxene, and feldspar.

Worldwide, the magmas that rose into various parts of the crust over the billions of years of earth history have produced varieties of minerals and rocks, including some rich enough in metal to be ore. The magmas moved within the crust as large, separate bodies of mineral crystals floating in hot liquids about 1100°C (2000°F). Some reached the surface to pour out as lava, but most hardened inside the crust, often below volcanoes. Major parts of mountain ranges such as the Rocky Mountains in Colorado and the Sierra Nevada in California were magmas that cooled at depth. These rocks and minerals are now at the surface in places because erosion has removed overlying crustal rock and movements in the crust pushed them up from the interior.

Over geologic time, some of the magma that hardened to rock within the earth now makes up mountain ranges such as the Sierra Nevada in California.

Many of the magma bodies that have invaded crustal rock contained more than elements that combine to form ordinary minerals, such as quartz and mica. Additionally, portions of the hot liquids carried ingredients for metallic ore deposits: atoms of copper, cobalt, manganese, tin, lead, zinc, silver, gold, chromium, and more. These metal-rich magmatic liquids are called hydrothermal solutions. They entered fractures in crustal rock, and even fractures in portions of the magma bodies that had already cooled. There, the hydrothermal solutions cooled, and the metal atoms within became concentrated in that part of the earth's crust. Where enough metal-bearing

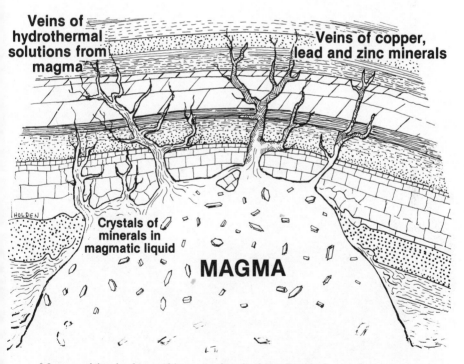

Magmas rising in the earth's crust often had hot liquids (hydrothermal solutions) that contained atoms of metals. Upon cooling, veins of metallic ore deposits formed in rocks of the crust.

minerals precipitated from the hydrothermal solutions, an ore deposit formed.

The area around Butte, Montana, became the "richest hill on earth" in this way. The Butte ore deposits owe their origin to the intrusion of a granite magma body tens of miles across that sent metal-rich solutions into a network of fractures in the crustal rock. When the fluids in this plumbing system cooled, they left "veins" of copper- and silver-bearing minerals named chalcocite, chalcopyrite, bornite, and tetrahedrite. Enough mineralized fracture-fillings formed at Butte so that since the 1880s miners have carved out a huge pit and 3000

The Butte, Montana, mining district. This view of the "richest hill on earth" is from around the year 1900. The Butte mines are no longer in operation.

miles (5000 km) of underground workings to reach ore in a region about the size of a major airport and two miles (3 km) deep. They mined over 3 billion dollars in copper as well as a wealth of silver, gold, and zinc from ore minerals associated with the copper.

Equally important for forming ore deposits is the ability of magma to heat and activate water already in the crust. This groundwater is derived largely from the atmosphere in the form of rainwater that seeped into the crust over millions of years. Because such subsurface water originated in the atmosphere, geologists refer to it as meteoric water.

Waters in the crust that were heated by upwelling magma also became hydrothermal solutions. Hot and circulating through rock fractures and pores, the solutions were able to leach, or remove, metal atoms from rock they passed through. This means that small amounts of metal atoms already in crustal rock were gathered up by the hot liquids, or hydrothermal solutions, that passed through channelways. The metal atoms were then carried off to accumulate elsewhere.

In 1962, geologists in Southern California discovered a striking example of this process. They were drilling in the Salton Sea area, a region that has hot springs caused by meteoric water heated by deep, hot rock and magma. Their goal was to reach a zone with water hot enough to provide geothermal power. Surprisingly, when the drill penetrated rock one mile down, the water there was not only over 300°C (540°F), but also rich in copper and silver atoms. This hydrothermal solution contained such high amounts of these metals that three months of pumping it to the surface coated the inside of the pipeline with 10 tons of copper and silver minerals.

Gravity, rather than hot solutions, has also created ore

deposits associated with magmas. One example developed 2 billion years ago, when magma intruded crust that is now South Africa. The molten rock occupied a large saucer-shaped chamber about 100 miles (160 km) across. As it cooled, chromite was among the earliest minerals to appear. These metallic crystals, usually only millimeters in size, were notably heavier than the hot magmatic liquid from which they crystallized. They therefore settled to the bottom of the magma chamber. Today this accumulation of minerals is an ore deposit consisting of twenty-nine layers of nearly pure chromite in the Bushveld Igneous Complex. The layers range from half an inch to 3 feet (1 m) thick and extend for tens of

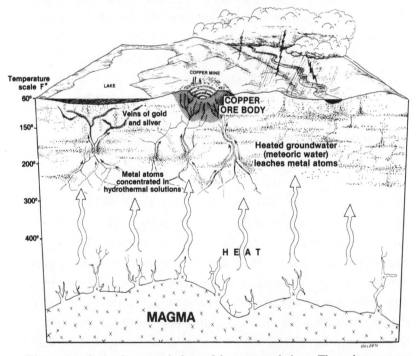

Water already in the crust is heated by magma below. These hot waters (hydrothermal solutions) circulate, leaching (removing) metal atoms from the rock they pass through. The metals are later deposited elsewhere as mineral veins.

miles. This ore-forming process, known as magmatic segregation, endowed South Africa with 80 percent of the world's chromium.

Magma adding rock and ore to the crust is a constructive process. But destructive weathering and erosion, which remove rock, create ore deposits, too. While the first process is "hot," using heat from the interior, the other is "cold," involving surface temperatures. However, like the hot process for forming ore, the cold one also depends heavily on water, as the formation of aluminum ore demonstrates.

Aluminum is the most abundant metal in the crust. Almost every rock has at least a few percent and some have over

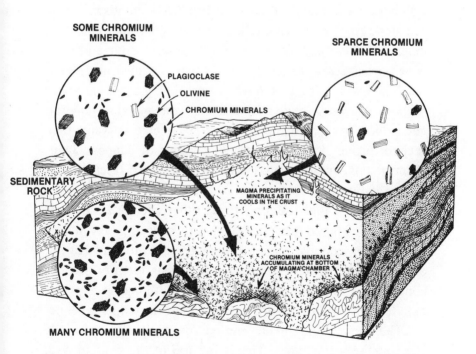

Minerals rich in metal, such as chromium, may settle to the base of a magma chamber due to gravitational attraction. There, they concentrate to form a metal ore deposit.

10 percent. There has never been a way, however, to profitably extract aluminum from ordinary rocks. But in nature, the continuous attack of moist, tropical conditions on certain aluminum-bearing rocks has converted at least some portions of the crust into ore. Where rainfall has been abundant and the weather hot, rainwater seeping through the rock leached practically all elements except aluminum and oxygen and carried them off. The residual aluminum-rich material became bauxite, a hydrous aluminum-oxide that is more like a hard soil than a mineral. Tropical regions, such as Jamaica, Guyana, and Suriname, have abundant bauxite and produce about 35 percent of the world's aluminum.

In Arizona, New Mexico, and Utah, another version of weathering and chemical reactions produced abundant copper ore deposits. Their origins began with the emplacement of magma bodies that contained only a little copper and iron, which occurred as scattered grains of sulfide minerals (metals combined with sulfur). Rainwater percolating downward over millions of years gradually dissolved the metallic minerals and added oxygen to their sulfur. The meteoric water then seeped farther downward, carrying atoms of copper, silver, other metals, and sulphate (sulfur plus oxygen). It left behind an "oxidation zone" of rock.

Underlying the oxidation zone, other chemical processes took place. The copper combined with oxygen in the waters to form, for example, minerals named malachite and azurite, or crystallized as native copper. Geologists identify this ore zone as one of "oxidized enrichment." Although it is profitable to mine, still richer ore formed below the oxidation zone.

The descending metal and sulfur solutions eventually reached the water table. This is the top of the zone of rock that contains permanent groundwater. There, with little oxygen in the environment, chemical conditions encouraged the metals, particularly copper, to combine with sulfur. The results are seen today as rich accumulations of metal-sulfide minerals, such as covellite and chalcocite.

These zones of "secondary enrichment," or "supergene" sulfide ore, are far richer in copper than almost any "hypogene" ore, the kind from hydrothermal solutions. At the water table, where secondary enrichment is greatest, ore can have over 2 percent copper. This compares with less than 1 percent in hypogene ore, and little or none in the leached rock overlying supergene ore.

During the formation of these copper deposits in the Southwest, leached metals did not move far from the source

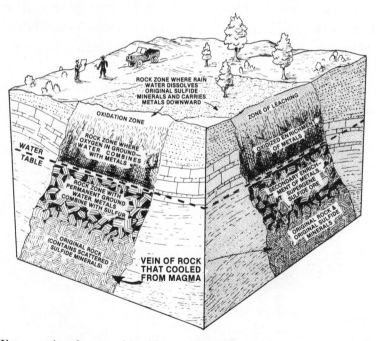

Water passing downward is able to redistribute small amounts of original metal, oxygen, and sulfur atoms in rock that cooled from magma. The results may include the formation of zones of new minerals richer in metal than in the original rock and suitable as ore. The process is called "secondary enrichment" and the ore is called "supergene," referring to its origin.

rocks above. They migrated downward only hundreds of feet. But other ore-bearing regions show that metals may travel great distances before reaching chemical and physical environments that enable them to precipitate as minerals. Metals may end up hundreds of miles from their source rocks.

In one example of when these conditions were met in the geologic past, the iron atoms flowed in streams to a shallow sea to form layers of iron ore large enough to blanket several states. The Clinton iron formation in Alabama marks the southern portion of that widespread ore-forming environment 400 million years ago along what was then the east coast of North America. About ten sedimentary iron-ore beds have been commercial in that region. Some are over 30 feet (10 m) thick, 9 miles (15 km) wide, and tens of miles long. They all contain ripple, wave, and mud features that indicate an origin where rivers met shallow seas.

Abundant oxygen is one cause of such iron concentrations. The chemical change from slightly acidic streams to nonacidic and more oxygenated seawater forced the iron atoms to precipitate as iron-oxide minerals. In a chemically

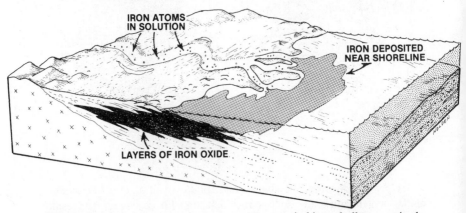

Where abundant iron atoms in stream water emptied into shallow seas in the geologic past, iron-oxide minerals sometimes formed as layers of rock that can now be mined.

similar way, manganese accumulates on the deep seafloor. The metal atoms enter the sea by rivers and submarine volcanoes. The products are walnut- to grapefruit-sized manganese-rich nodules scattered on parts of the seafloor as though dumped from a passing ship.

For some metal deposits, such as uranium ore in the Southwest, bacteria played a key role. Buried in sandstones and thriving on sulfates in groundwater, certain bacteria released sulfur as a by-product. When uranium atoms moving in groundwater in the sandstones entered the sulfur-rich environments, they became insoluble and crystallized to ore minerals.

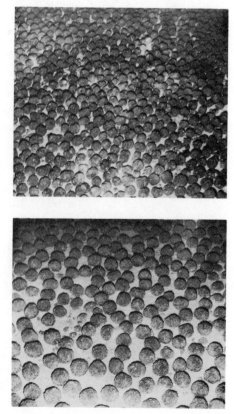

Nodules comprised mainly of manganese and lesser amounts of nickel, cobalt, and copper that formed on the Pacific Ocean floor.

Ore deposits do not always need solutions to form. Sometimes ore formation is purely mechanical, as when minerals are transported by streams. Although weathering attacks all material near the surface, some minerals are too durable to be affected greatly. After falling out of aging, crumbling rocks, they remain as tiny, hard grains that end up in streams and rivers and on beaches. If they accumulate in amounts worth mining, they are "placer" deposits (placer rhymes with gasser). This is true of gold—the mineral that more than any other has added excitement and color to finding ore deposits.

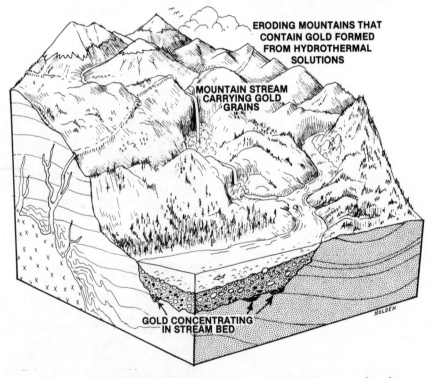

Where metal-rich resistant minerals, such as gold, weather out of rocks, streams may carry off the grains along with sand and gravel. In areas of the stream where the water is slow-moving, the gold grains may settle out of the water. Large accumulations of water-deposited metallic minerals form placer ore deposits.

The great Alaskan gold rush was for placer gold. It began in the Klondike, Northwest Territories, Canada, on August 16, 1896, when three hunters discovered gold in a stream that flows into the Yukon River. They recovered it by panning, or using flat pans to slowly sift and wash the stream sediments. This method separated the grains of heavy gold from the lighter minerals like quartz, feldspar, and clay. In days, the three men panned nearly 9 pounds (4 kg) of gold. That would be worth about 65 thousand dollars today. They appropriately named the stream Bonanza Creek.

Word of the gold strike spread. Within a year, ten thousand men from all parts of the world set out for the Klondike. In two years, the town of Dawson grew from only a few houses to over twenty thousand residents. Some prospectors struck it rich quickly. One from San Francisco found 150,000 dollars worth of gold in less than a month. Others had to pan Klondike streams for a year before recovering 100,000 dollars worth. But thousands of fortune seekers never found gold,

Gold prospectors of the 1800s panning a stream for gold flakes and grains.

and many others lost their lives to cold winters, diseases, and guns. In all, the Klondike miners found over 100 million dollars in placer gold before the supply began to dwindle in 1906.

Practically the same situation exists today in Brazil, where 250,000 *garimpeiros,* or prospectors, are scouring the Amazon region for gold in the gravels of the Rio Madeira and in the soil of a great pit called Serra Pelada. Once a mountain, Serra Pelada has been gradually carved away by the thousands of *garimpeiros* swarming the land like ants. For some, the labor of hauling away soil in plastic bags to sift and examine paid off in the discovery of giant gold nuggets. In 1983, for instance, one discovery reportedly weighed 137 pounds (50 kg), worth over half a million dollars.

While these placer gold deposits formed by mechanical surface processes, namely stream-flow carrying heavy minerals, the origin of the individual gold grains was different. This native mineral is magmatic. That is, it crystallizes from hydrothermal solutions associated with magmas. This is true of most metallic minerals. An important question, then, for understanding ore deposits is: Why were there magmas to form minerals and hydrothermal solutions in the first place? Examining the architecture of the earth answers this question and explains why many ore deposits formed where they did.

3

Where on Earth . . . ?

The minisubmarine *Alvin* was nearly 2 miles (3 km) below the surface of the Pacific Ocean, inching along the seabed and piercing the darkness with outboard spotlights. Inside, Bill Normark, a U.S. Geological Survey scientist, searched the ocean floor for metallic ore deposits and for a better understanding of how they form. His investigation came after reading oceanographic and geologic reports written in the 1970s about mounds of sulfide minerals on the seafloor containing iron, zinc, nickel, and manganese.

One report was written by scientists who dived in the French submarine *Cyana*. They described a zone of the southeastern Pacific floor covered with sphalerite, the same mineral that is mined on land for zinc. They also reported abundant pyrite, an iron-sulfide mineral, and mixtures of gold, silver, lead, nickel, and copper forming various other minerals. The deposits occurred in a "field" of seafloor mounds about 30 feet (10 m) high, half that distance across, and spaced 10 feet (3 m) apart.

In another report, scientists on an earlier *Alvin* expedition had found fractures in the ocean crust that spewed streams of warm water into the cold Pacific Ocean. While normal seawater at depth can be as cold as 2°C (35°F), the balmy water at these volcano-like vents measured 11°C (52°F). Sea creatures never observed elsewhere thrived in sunless colonies flushed by these warm waters. The warm seafloor vents offered the scientists a fantasy world of monstrous clams, crabs, and tube worms, plus varieties of bacteria living on the chemicals spilling from the spouts of water.

Normark's *Alvin* dive was equally rewarding. Most notable was the discovery of a "pipe," or rock chimney, furiously discharging a plume of black water. Using *Alvin*'s robotic arms, Normark knocked off part of the 15-foot (6-m) "smoking" chimney and placed a thermometer inside. The water temperature, however, was too high for the instrument to record. The hot water melted the plastic. This indicated a temperature greater than 315°C (600°F). The black color and

By diving in the mini-submarine *Alvin*, scientists have discovered ore deposits forming on the seafloor.

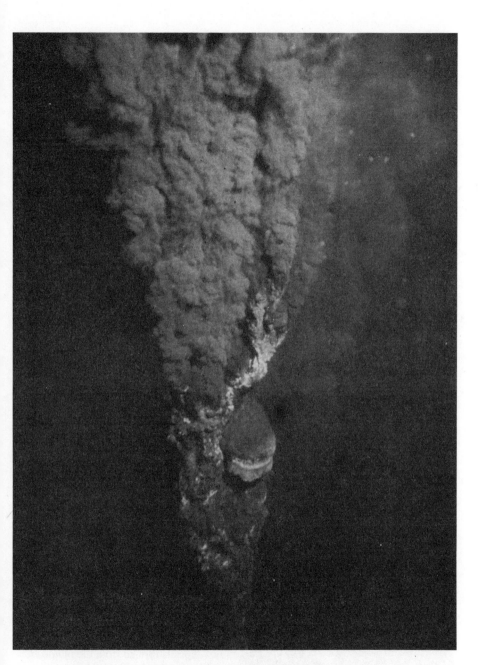

A rock chimney, or "smoker," on the Pacific Ocean floor spewing hot metal-rich water (hydrothermal solutions) into the seawater.

temperature, as determined by later studies of water samples and temperatures, was caused by the large amounts of metal atoms in water that was 380°C (715°F). Normark found what he had sought: hydrothermal solutions streaming from crustal fractures at the ocean bottom and transporting metals to deposit in and on nearby rock.

Since then, other hydrothermal vents, or "smokers," have been located in that region of the Pacific as well as off the coast of Washington state. Actual underwater ore deposits have been discovered, too. For instance, scientists on an *Alvin* dive in the region of the Galápagos Islands in 1981 happened upon a glimmering body of copper and zinc ore large enough to fill a baseball stadium. Communities of giant dead clams draped the deposit. The way they hung from rock towers and chimneys that had long ceased bellowing hot fluids brought to mind a seafloor holocaust.

A "smoker" of hydrothermal fluids blasting out of a vent along the East Pacific Rise, 21° north latitude. A portion of the rock "chimney" is behind the smoker. Other rock is mineral deposits formed from the plume of hot water. Deep-sea fish and crabs are on the deposits.

It came as no surprise to geologists that when ore deposits were finally observed in the making, the southeastern Pacific Ocean floor would be the place. That region has a geologic feature called the East Pacific Rise. This is a submarine mountain range of volcanoes and deep fractures in the crust. Magmas erupt from this fracture system to construct the "rise," add rock to the crust, and create hydrothermal solutions out of seawater that enters the fractures.

The East Pacific Rise is part of a nearly continuous 40,000-mile (65,000-km) mountain range of ocean-floor volcanoes spanning the globe. Not all parts of this midocean ridge system, as it is called, are creating new rock from rising magmas and spewing hot solutions at the same time. In fact, such activity may occur at a particular place only once in thousands of years, just as volcanoes on land may lie dormant for long periods before coming to life again. But the 4½ billion years that the earth has existed and the length of the midocean ridge system assures that ample deposits of ore have formed or are now forming on the seafloor.

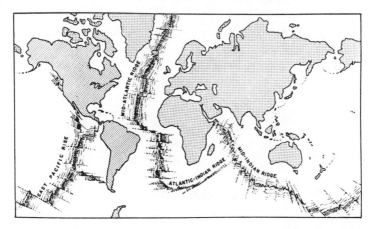

The midocean ridge system of the earth—a mountain range made largely of volcanoes. Ore deposits have been observed forming on the portion of the ridge called the East Pacific Rise.

The midocean ridge system originated because the earth's outer shell, or lithosphere, is broken into "plates." As the plates move apart, a few inches each year, fractures form, magmas rise, and volcanoes grow. Plates are about 60 miles (100 km) thick, and their uppermost material is either continental or oceanic crust.

Even before the discovery of smokers on the East Pacific Rise, ore formation was suspected in the Red Sea. There, too, plates are moving apart, creating a spreading seafloor. Africa and Saudi Arabia are gradually separating, and magmas lie below the "spreading center." Seawater seeping into fractures is warmed and later escapes upward to the seabed. It rises with metal atoms scavenged from the rocks through which it passes.

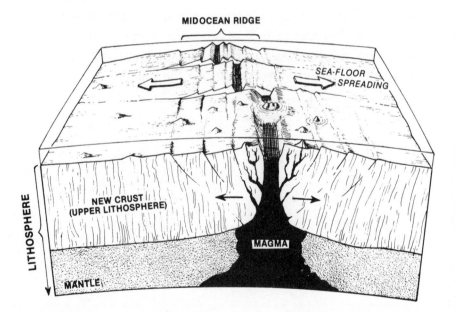

At a midocean ridge, where lithospheric plates spread apart, magma rises to cool to form new crust. Volcanoes and hydrothermal solutions form during the process.

High water temperatures and salt content in the Red Sea attracted scientists' attention as far back as the 1880s. For long afterwards, the sea was thought to be unusually warm because of its tropical location, and saltier than normal because evaporation concentrated its salt content. But in 1966, oceanographers aboard the *Atlantis II* on a cruise through the Red Sea to reach the Indian Ocean made a one-day stop for samples of the salty water and ooze, or seafloor mud. They were interested in a particularly deep part of the sea. This was a depression about 2 miles (3 km) across and more than 6000 feet (1830 km) deep and called the Discovery Deep. During sampling, however, the ship drifted, and the scientists accidentally located an even larger "deep," about 5 miles (8 km) across. To their surprise, they sampled water at 56°C (133°F),

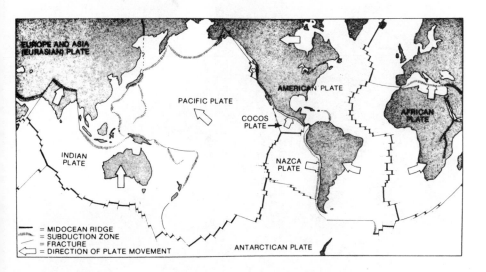

The earth's outer shell, or lithosphere, is made of several large "plates" that slowly move a few inches per year. Spreading centers, such as the East Pacific Rise and the Mid-Atlantic Ridge, form where the plates move apart. Subduction zones form where plates collide.

black ooze that was too hot to touch, and water and ooze thick with iron, manganese, zinc, and copper. They named their discovery the Atlantis II Deep.

Seawater from the Atlantis II Deep is 7 times saltier than ordinary seawater. It contains 30,000 times the lead, 25,000 times the manganese, 5000 times the iron, and 500 times the zinc of ordinary seawater. If the Atlantis II Deep could be mined as an ore deposit, its contents would be worth billions of dollars.

The Red Sea ore formation differs from that occurring in the East Pacific Rise only in that the environment is not an

Scientists aboard ship examining metal-rich sediments cored from the bottom of the Red Sea.

open ocean. Rather, the Red Sea is an ocean in the earliest stages of development. It is slowly expanding as continental land mass breaks apart and magma fills the space between to cool and create ocean floor. Geologists must conclude, then, that metal deposits can form in all stages of ocean development, from that of small sea to that of full-sized ocean. All that is needed is a heat source and fluids to circulate throughout fractured rock.

Because ocean crust is continually created and the earth is not increasing in size, crust must be destroyed somewhere. Where the floor of the eastern Pacific Ocean meets the South American plate, it is forced downward, or subducted, beneath

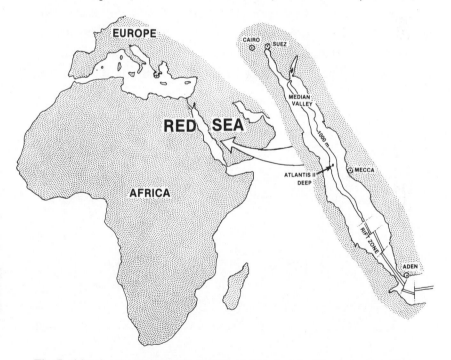

The Red Sea is spreading apart, marking the beginning of a new ocean. Heat from magmas below the rift zone in the center of the Red Sea have created hydrothermal solutions from the seawater which, in turn, produced metal-rich seafloor "ooze."

43

South America. The northern Pacific Ocean seafloor moves north to become subducted below Alaska. In both regions, and in many more worldwide, old ocean floor (and an entire lithospheric plate) moves down into the earth's warm interior. Once deep in a subduction zone, the old crust partially melts and forms magmas that rise.

Subduction zones form where two plates collide and one overides the other. Because magmas form at these plate boundaries, subduction zones are sites of ore formation, like the boundaries between two separating plates at midocean ridges.

In the western regions of Central and South America, ore deposits are tied to magmas rising from subduction zones. Numerous large bodies, or plutons, of granitic rock emplaced in the crust provided the sources for copper in Mexico, Ecuador, Peru, and Chile. Although some of the mining districts owe

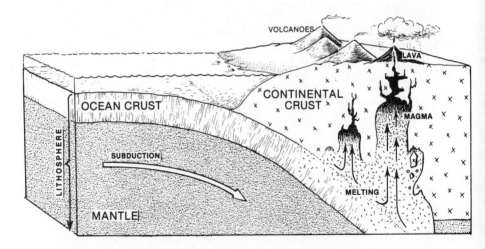

Where plates collide, a subduction zone may form as one plate is forced beneath the other. The subduction process causes melting and the formation of magmas that rise above the subduction zone. Hydrothermal solutions associated with the magmas may produce metal ore deposits.

their existence to supergene enrichment, it took plutons of magma to originally provide the regions with the atoms of metals such as copper.

The states of Arizona, New Mexico, and Utah have nearly one hundred copper and molybdenum deposits associated with plutons. Geologists cannot demonstrate that every occurrence of copper or molybdenum is linked to subduction processes because some areas show no evidence of present or past subduction zones. But the subduction and magma-forming processes active 50 to 100 million years ago along western North America do explain the presence of many Southwestern ores.

Collisions of lithospheric plates have sometimes pushed oceanic crust up and onto continental crust. In some regions, this obduction, as it is called, delivered metal deposits from former seafloor spreading centers to land. Geologists call bodies of such upthrust ocean floor ophiolites. Mankind's earliest copper mines are in ophiolites on the island of Cyprus in the Mediterranean Sea. This region was mined by early Greeks and Romans, and is now known to have about ninety individual copper-iron-zinc deposits. In fact, the word "copper" was derived from Kypros, an old name for this island.

Plate boundaries such as spreading centers and subduction zones are geologically active areas of the earth and are therefore likely places for ore formation. But ore has also formed in the interiors of lithospheric plates. One large occurrence of ore in an intraplate region is the tri-state district of Kansas, Missouri, Oklahoma. It contains mainly lead and zinc minerals in sedimentary rocks. Curiously, the fluids that formed the ore were hydrothermal, but there is no geologic evidence for a midocean ridge or any volcanic vents having been within hundreds of miles of the tri-state region. Equally puzzling is that the sedimentary rocks containing the minerals formed in shallow seawater, such as near an ancient shoreline, several hundred million years ago. Geologists conclude, then, that hydrothermal fluids must be able to migrate great distances, tens or even hundreds of miles from ocean-floor vents, before reaching the proper chemical environment to deposit metals.

The geologic study pertaining to movements of lithospheric plates is called plate tectonics. Understanding plate tectonics, or where present and former plate boundaries are located, helps geologists determine how and where ore deposits formed. Yet there is one process that created large metal deposits in areas like Sudbury, Ontario, Canada, by forces from outside the earth.

Sudbury is one of the world's largest suppliers of nickel. The mineralized rock formation is elliptical in outline at the surface, about 40 miles (65 km) across in its largest direction, and funnel-shaped at depth. Sudbury ore originated between 1½ and 2 billion years ago.

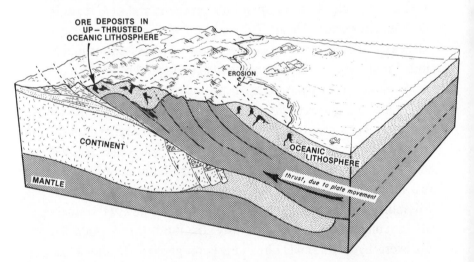

Some plate collisions pushed, or thrusted, oceanic lithosphere onto continental crust. Because the slice of ocean crust had formed at a midocean ridge, hydrothermal ore deposits may be brought onto land. The process is called obduction, and the slice of oceanic lithosphere is called ophiolite.

Until the 1950s, geologists considered Sudbury to be another example of mineralization due to magma that upwelled into the crust and cooled. But around that time, Robert S. Dietz, a geologist with the United States government, began a detailed examination of the rocks around the Sudbury deposit and made what seemed to be an outlandish interpretation. His observations pointed to a blast of energy equivalent to that of several nuclear bombs having hit that region of Canada in the geologic past. He found that the rocks contain shock features, such as glass created when minerals were destroyed by a sudden pressure increase equal to thousands of tons. Dietz also found minerals that can form only under high pressure or shock conditions, such as a type of quartz that requires at least 150 tons of pressure per square inch to form. And certain Sudbury rocks were fragmented as though crushed and granulated by a sudden blow.

The conclusion Dietz reached was that the Sudbury region was struck by a giant meteorite, a large body falling to earth from space, about 2 billion years ago. The meteorite was 2½ miles (4 km) across and traveling at 10 miles (16 km) per second. Its impact carried enough energy not only to shock and partially melt the crust but to excavate an enormous crater in it. It did not, however, deposit the ore.

Instead, the removal of rock by the blast reduced pressure on the underlying lower crustal rock. This "pressure release" enabled instant melting, or magma formation, deep in the lithospheric plate. The magma rose to fill the "explosion" crater and surrounding crustal fractures. Once magma was present, ore formed from the associated hydrothermal fluids. The Sudbury nickel deposit differs, then, from other hydrothermal deposits because of the origin of the magma necessary for the metallic minerals. It originated because of the impact of a large iron meteorite—an external force—and not from internal processes, as magma almost always does.

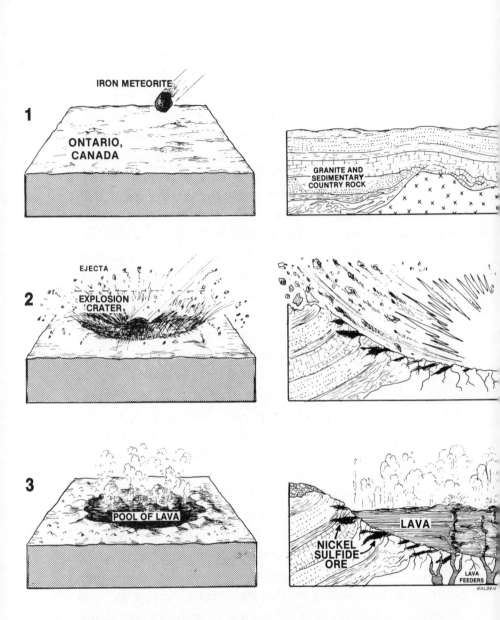

1 IRON METEORITE
ONTARIO, CANADA
GRANITE AND SEDIMENTARY COUNTRY ROCK

2 EJECTA
EXPLOSION CRATER

3 POOL OF LAVA
NICKEL SULFIDE ORE
LAVA
LAVA FEEDERS

Meteorite impact may have produced some ore deposits. The reason is that the impact may have exploded out enough crustal rock to enable instantaneous melting, or magma formation, at depth (due to pressure release). Once magma formed, hydrothermal solutions created ore deposits. This may have been the origin of the Sudbury, Ontario, metal deposits.

Whether the origin of a metal deposit is due to plate tectonics, meteorite impact, or to the accumulation of metals after weathering and erosion, the metal must be discovered before becoming useful to mankind. Geologists and mining engineers use a variety of exploration techniques to find those deposits for the manufacture of metal products useful in society.

Underground mining of metals, such as nickel, at Sudbury, Ontario.

4

Finding the Riches

In 1959, Forbes Wilson was a geologist working in Indonesia for the Freeport Sulphur and East Borneo companies. His mission was to find a copper deposit that the two companies could mine in partnership. The job required crossing miles of mountains by foot and jeep. Therefore, before geologist Wilson set out, he did library research to save time and work exploring in the mountains. By searching through geologic reports on Indonesia, he hoped to find clues about places likely to have ore.

Wilson's library work proved valuable. Twenty-three years earlier, two petroleum geologists on a mountaineering holiday in New Guinea had spotted signs of copper on Mount Carstensz. They later mentioned their observation in a 1939 scientific report published by the University of Leiden in The Netherlands. Wilson discovered that report during his literature search. He went to Mount Carstensz in 1960 and rediscovered a huge "knob" of copper ore now called the Ertsberg. Production of 7400 tons of copper ore per day began there in 1972.

Most of the earth's surface has been visited and inspected. Therefore, the first step in exploring a new region for metal is to search through the geologic literature about that region. Computers make these searches a fast, low-cost way to explore an area for ore potential without actually visiting it. In most cases, however, reports provide only general scientific information. They seldom lead exploration geologists directly to ore bodies, as in the case of the Ertsberg. This is mainly because ore deposits that are relatively easy to find at the surface have already been discovered. Finding a new metal deposit requires exploration surveys that use teams of people to examine land that they initially learned about by reading scientific literature.

The first kind of survey is usually general, such as covering a large tract of land by air or jeep or even on horseback. The amount of land covered in a general, or reconnaissance, survey can be the size of a mountain range. The second type of survey is a detailed inspection. It covers the areas that reconnaissance showed most likely to contain metal deposits.

Exploring by air does not necessarily require flying by plane or helicopter. Geologic information for most areas is already available in collections of photographs taken from airplanes and satellites. Aerial photographs and satellite images help identify various rock types, crustal fractures, vegetation, and ground temperatures. These features guide geologists to specific locations to prospect for metal.

The application of satellite images to ore exploration is called remote sensing. It is particularly useful in areas that are difficult and costly to reach. In Greenland, for instance, a 1984 study used multispectral scanners orbiting the earth in Landsat satellites to detect color differences on the ground. The scanners measured the wavelengths of solar radiation reflected from the different ground features—rock, soil, and

A general, or reconnaissance, survey of land is commonly the first step toward locating ore deposits.

After reconnaissance geologists select certain rocks to examine in detail.

Photographs of a region taken from an airplane provide valuable information during reconnaissance surveys of a region. Here, a geologist compares aerial photographs with actual features on the ground.

This satellite photograph illustrates geologic features in Grand Teton National Park, Idaho. It was taken from the Earth Resources Technology Satellite-1 (ERTS-1).

vegetation. Computers helped convert the measurements into color differences that could not otherwise be detected. Wherever orange-brown, or "rusty," colors showed on the images, abundant iron must be present. That remote-sensing project located eighty-eight such places in east central Greenland that will eventually be field-checked for iron-sulfide mineral deposits that may also contain nickel, copper, and zinc.

What photographs or satellites do not show can perhaps be learned from scientific instruments taken aboard an airplane or helicopter. Measurements are made while the aircraft sweeps back and forth across a targeted area. Airborne surveys measure the magnetic, electrical, reflective and radioactive properties of the rocks below. For example, aeromagnetic and electromagnetic aerial surveys can locate metal-sulfide deposits because the instruments detect high amounts of iron in the rocks below. Airborne imaging spectroscopy identifies minerals by measuring their light-reflectance. And gamma-ray detectors in helicopters flying 50 to 75 feet (15 to 23 m) above the ground can pick out rock formations that have radioactive uranium. All these measuring techniques use physical properties of the earth, and their application to exploring for ore is therefore called geophysical exploration.

This DeHavilland Twin Otter does surveys using instruments that detect electrical and magnetic properties of the earth's surface.

After air surveys locate a region that may have ore, the same geophysical methods are used by ground survey teams to explore further. Ground crews can also search by checking the gravitational attraction of a region. Gravity meters locate concentrations of heavy elements, such as iron and lead. Electricity applied to the ground can find ore. This technique was one of the earliest that geophysicists used. It began when the French geophysicist Conrad Schlumberger located a new copper deposit in the Bor mining district of Yugoslavia in 1914. He planted poles in the ground, applied electricity, and measured the electrical current. Schlumberger's achievements in ore exploration led to the application of these geophysical techniques to petroleum exploration in the 1920s. A testimony to its success is that today the world's largest oil-well-servicing company bears Schlumberger's name.

A magnetometer is an instrument that evaluates the magnetic properties of the rocks in a particular region.

Gravity meters on this snowfield determine the gravitational properties of the rocks below.

Setting up wire lines are part of the electrical surveys used in exploration for ore deposits.

Electrical surveys for ore deposits can be made by instruments carried aboard helicopters.

Chemistry also contributes to metal exploration programs, especially where ore bodies are expected to be small and hidden by rock and soil cover. Because soil is weathered rock and because streams carry dissolved metals, chemical tests on soil, stream water, and stream sediment may reveal evidence of a nearby metal deposit. In the Platreef nickel-platinum deposit of South Africa, soil less than 6 feet (2 m) thick carries high amounts of nickel, cobalt, copper, iron, mercury, and arsenic from the ore body below. But where the soil is thicker, only the presence of mercury reveals the location of an underground ore deposit. The reason is that mercury travels upward from the buried ore as a vapor and can be detected no matter what the soil cover.

Mercury-vapor detection has had other uses. One example pertains to Chinese history. The Chinese believe that the mausoleum constructed for Emperor Qin Shi Huang about 200 B.C. contains a buried reservoir of mercury and mechanical devices to circulate it. Supposedly the Emperor was buried with mercury, a metal that is liquid in its normal state, to symbolize surface waters, such as rivers and seas. Archaeolo-

Sampling and analyzing stream water can provide clues to the metals in the nearby rocks.

gists, eager to determine whether this legend was true, but without digging and destroying the tomb, analyzed the overlying soil for mercury. Interestingly, they found nearly ten times more mercury in the soil above the tomb than in soil surrounding the mausoleum.

Examination of vegetation also reveals the presence of metals. Tree roots feeding on metal-rich soils may carry clues of buried metal deposits to branches and leaves. Chemical analyses of vegetation may therefore substitute for analyses of soil and rock that cannot be sampled easily. In one study, U.S. Geological Survey geochemists investigated the possibility that pine needles give clues to gold ore below.

Another type of exploration that combines geochemistry and geobotany is to search for vegetation that thrives on certain metals. Uranium was discovered in the Yellow Cat area of Utah in this way. Several types of plants, such as the aster, thrive on selenium, an element associated with uranium ore. Of dozens of areas covered by selenium-indicating plants, 46 percent proved to have uranium-mineralized rock within 200 feet (70 m) of the surface.

Portable geiger counters help establish whether or not uranium is present in rocks.

No matter how the geologic, geophysical, or geochemical information is collected, it is of little value until sketched on a map or aerial photograph. This allows exploration teams to study patterns of ground magnetism, for instance, or metal abundances in soil and water for their promise of leading directly to ore. Equally important on a map are outlines showing where the different rock types occur in the exploration region, and lines marking ancient crustal movements, or faults. Such geologic maps, with geophysical and geochemical maps, generally lead to the final exploratory step—drilling to obtain samples of rock below the surface.

A diamond-studded drill bit at the end of a string of drill pipe about 2 inches (5 cm) in diameter can bring up thin cores of rock from as much as a mile (1.6 km) below the surface. Examination and analysis of rock cores reveals the amount and kind of metal present and the exact depth at which it occurs. Costs of drilling are high, as much as 25 dollars a foot. But exploratory drilling must be done before attempting the most expensive part of obtaining ore. This is the actual mining.

A U.S. government geologist discusses a geologic map of a region in New Mexico. Geologic maps show what rock types are present at various locations.

Drilling and coring provides samples of the rocks below the surface. Geologists study the cores to determine the likelihood of ore deposits.

A close-up view of a drill bit used in exploration drilling for ore deposits.

Determining how to mine ore depends greatly on what drill-core samples indicate about the shape and depth of the ore body. If the body is shallow and broad, an open-pit mine is most efficient. Mining engineers develop an open pit as though they were carving a bowl-shaped football stadium out of the earth. The deepest part of the mine is near the center, and the walls of the pit go up in a series of steps, or benches. The benches spiral upward from the center to the top so that they serve as a roadway for ore-hauling trucks. Engineers mine and expand the site by drilling vertical holes in the faces of the benches, filling them with explosives, and blasting them away.

Open-pit mining for copper in southern Arizona.

The world's largest open pit is The Hill, a copper mine at Bingham Canyon, Utah. An entire mountain has been cut away to leave a pit 2 miles (3.2 km) long, 1½ miles (2.4 km) wide, and slightly over half a mile (0.8 km) deep. In full operation, enough rock can be removed each day from the Bingham pit by railcars and trucks to fill up a basketball arena.

If an ore body is deep and narrow, vertical shafts and horizontal tunnels, or drifts, are carved into rock to reach the ore. Mining underground is more complex than it is in an open pit. It requires railcars, elevators, or conveyor belts to carry out the broken rock. Also needed are pumps to remove groundwater that seeps in, timbers to support the walls, and fans to circulate air. A common method of removing ore is "stoping." This involves drilling and blasting the ceilings and walls and letting the broken rock accumulate for later removal.

Heavy machinery removes rock in the open-pit copper operation in Baghdad, Arizona.

Much of the world's metal comes from ore deposits mined from shafts and tunnels.

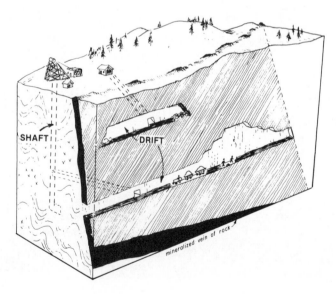

Vertical and horizontal tunnels in underground mines are called shafts and drifts, respectively.

Ancient miners also took ore from underground mines. They removed water that seeped in by hauling leather bags or wooden buckets, and they broke the rock with hammers and wedges. Not having drills and explosives, they sometimes built fires against the mine walls to heat and crack the rock.

Future mines may include those that produce metal from underwater deposits. Likely places will be along midocean ridges, where copper, zinc, cobalt, manganese, and nickel minerals accumulate from the hydrothermal solutions. Manganese nodules may eventually be sought by mining companies and scooped up or vacuumed from the sea floor. Any open-sea mining, however, will involve political problems about which country is entitled to reap profits from metals recovered. As yet, there are no mining laws of the sea that satisfy everyone.

Wherever deposits are mined—mountain range, desert floor, or sea—metals must be converted to the many products people want. How, then, do metals come into use and exactly what are the needs for the many different metals?

An early underground miner.

5

Putting Metal to Work

Early in 1980, politicians and business people became deeply concerned about the economy of the United States. During the last two months of 1979, Iran took ninety hostages, most of whom were Americans, and Soviet military troops invaded Afghanistan. The world faced new crises and uncertainties about international economy and supplies of oil from Middle Eastern countries. Worries about inflation, the increasing costs of goods, overflowed into the banking business and the metals market. Many individuals decided that the best way to protect themselves against a possible worldwide financial crisis was to own gold and silver.

In September 1979, gold was selling for about 400 dollars an ounce, and silver cost about 12 dollars an ounce. Those prices seemed low, and people seeking economic safety from world problems began to buy heavily in the last days of 1979. On January 2, 1980, the demand for gold pushed it to 500 dollars an ounce, and silver soared to 40 dollars. In following days, gold moved in leaps of 20 to 50 dollars a day. Those who wanted to make a fast profit sold. Those who believed that these metals were still selling cheap bought. The demand for gold and silver grew more than the desire to sell the metals, and prices swelled. Between January and March, over 150 tons of gold in jewelry and bars were bought and sold from private stocks. Each day on the floor of the Commodities Exchange Center in New York City was wild with metals traders buying and selling contracts around the world.

The demand grew to such a point in January 1980 that gold rose to over 800 dollars an ounce and silver to over 50 dollars. Those who bought low and later sold at these lofty prices made fortunes. They were wise to sell, because by spring the demand for gold and silver had diminished, and prices began to slip. Speculators who missed the low prices and had bought gold at premium prices, believing they would

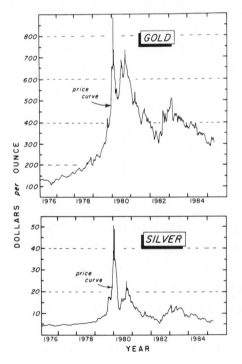

The up and down changes in prices of gold and silver over the period 1976 to 1985. The highest prices reached by these metals occured early in 1980.

OPPER   COMEX SILVER   COMEX GOLD

Metal traders on the floor of the Commodity Exchange in New York City. Most of the numbers listed on the board are prices that traders are willing to pay or sell in future months. As commodities, metals are sold by ounces or pounds.

go even higher, lost money. By May 1980, these metals were down to 500 dollars an ounce for gold and 12 dollars for silver.

Demand for metal is one of the main ways to establish its price. Whether or not there is demand depends on how useful the metal is in society. Gold and silver, for example, have applications in electronics, dentistry, jewelry, and photography. But in the case of the 1980 panic to buy these metals, they were in demand because of their history of retaining value during uncertain times for world economy. For gold, this quality stems from its having been the original standard on which money was based.

Although gold is no longer the standard for money, people still value it as protection against inflation. Many believe that gold's price will increase enough to offset any increasing costs of food, household goods, and services. Gold will always be valuable, no matter what world problems exist. The same is true of all precious metals, which include silver and members of the platinum group, such as palladium. But gold is the principal and most publicized precious metal.

The supply of a metal is the second main way to establish its price. In the case of copper, supply is abundant because many African, North and South American, and Philippine mines produce it. The price is therefore relatively low, about one dollar a pound. Even though copper has long been an important commodity, used largely as electrical wire and plumbing pipe, the demand is not great enough to increase its price.

How people use metals such as iron, copper, aluminum, gold, and silver is evident from the buildings, automobiles, telephone wires, and food containers that have become essential in modern society. Not obvious, however, are the uses for most other metals. Lead, for example, is used in batteries. Zinc is alloyed with copper to make brass, and is also used to galvanize, or coat, steel to prevent rusting. Tin is used in solder, to plate other metals, and in alloys such as bronze. These three metals, along with copper, make up the nonferrous group of metals.

Iron, manganese, nickel, chromium, tungsten, vanadium, and cobalt belong to the ferro and ferro-alloy group. All of them are used in the production of steel. Vanadium, for example, included with iron when making steel, gives the resulting alloy extra strength.

The booming space and computer industries have created strong demands for the not-so-common metals such as germanium, zirconium, beryllium, rhenium, and rhodium. These metals and about twenty more are vital components in defense weapons such as lasars and warheads, and in aircraft, satellites, and the space shuttle program. If supplies of these metals were cut off, advances in electronics, computers, and military defense would halt. Therefore, certain metals are considered "strategic." That is, they are critical for national defense and scientific development.

Strategic metals are a political issue. For instance, most of the world's supply of vanadium, manganese, chromium, and platinum comes from only two countries, South Africa and the Soviet Union. The United States and other nations therefore take precautions against political disputes that could lead to cutting off supplies of these strategic metals. They accumulate, or stockpile, at least a year's supply.

Each nation's stockpile of strategic metals differs, however, depending on how easy or hard it is to import a particular metal. While West Germany stockpiles cobalt to use as a superalloy in turbine engines and cutting tools, Canada does not have to. It produces cobalt at Sudbury.

Getting metal from the ground and into useful products requires various processes to concentrate it. In almost every case, the rock containing metal is crushed and ground to particles about the size of peas. Crushing is done at the mine to eliminate costs for shipping both the metal and the boulders of worthless rock containing the metal elsewhere for treatment. The crushed ore goes to a mill. There it is put into a revolving steel drum containing steel balls that crush it further into a "flour."

Crushing and milling are mechanical operations. To separate the metal-bearing flour from the powder of ordinary rock

Among the steps taken to extract metal from rock is to crush the rock into small particles.

A mill superintendent watches over the rock crushing process.

Transporting ore from a Colorado molybdenum mine to the mill for further crushing.

These huge mills turn the rock fragments that contain metal until the rock and metal minerals it contains are "flour."

generally requires chemical processes. Flotation is one method. When ore powder is placed into various types of liquids, the metal powder will either sink or float away from the rock powder. To assist the process, air is sometimes blown through the liquid to form a bubble froth. The bubble walls may attract the metal, and the froth can then be skimmed off to concentrate it. The metal concentrate is melted at a nearby smelter and poured into molds to cool. This smelting process further separates metal from the remaining rock powder.

Flotation troughs that use liquid chemicals to separate metal from rock powder.

Mining, crushing, milling, and smelting have gone on for centuries. Smelting, for example, was probably discovered when early man unknowingly placed lumps of ore around his fire and then noticed metal melting due to the heat. The long history of extracting metal from rock indicates that much of the earth's supply of ore is used up. Will the effect of the steady depletion of metal deposits affect future generations?

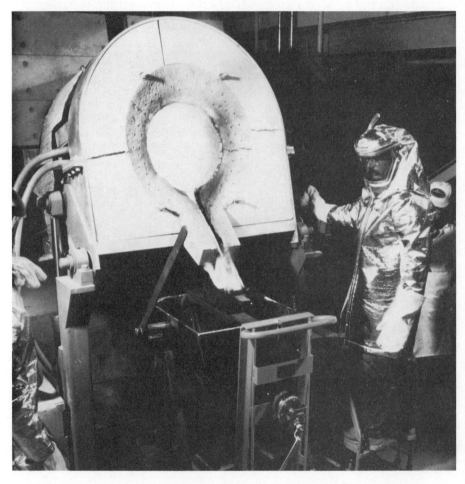

Pouring molten gold at the McLaughlin Mine, California. Temperatures in this high-security "gold room" reach 1093° C (2000° F).

Concentrates of metal are melted and poured into molds at a smelter near the Sudbury nickel mines in Ontario.

6

Caution: Supplies Can Run Out

The Vuonos copper mine at Outokumpu, Finland, had been in operation only thirteen years when management announced in 1985 that operations were shutting down. After 5 million tons of copper, zinc, and cobalt ore had been extracted and processed, the ore in that part of the crust had run out. Vuonos was no longer a metal reserve, or a region that could supply metal.

Metal reserves have been exhausted, or used up, in many areas throughout mining history. The Mother Lode, which was the ore deposit at the center of the 1850s gold rush in California, is another example. Most of the gold there was gone after about ten years of continuous mining by frantic gold-seekers.

The terms "reserves" and "resources" are commonly used when evaluating how much ore remains to mine at today's costs and technology compared to how much may be available in the future. Reserves are deposits known to contain enough metal to be worth mining under present economic conditions and technology. Drilling near an existing open-pit mine, for example, may show that ore is present for a half mile (1 km) beyond the pit boundary. The mine operators therefore know that their reserves are larger than the present mine site and that they can later expand the mine to stay in business.

On the other hand, metal resources are areas where metals exist in the earth in more than normal amounts. Accordingly, resources can be plentiful. An iron-rich soil, for instance, might be a resource. But there are unlikely to be large enough quantities of iron to allow setting up a mining operation. That is to say, the grade, or the percentage of metal in the rock and soil, is low. If new technology suddenly enables a resource to be mined, the deposit becomes a reserve. For example, rocks had to contain 13 percent copper to qualify as ore in 1700, but 2½ to 5 percent copper was adequate grade in 1900. Today, ½ percent grade copper can be mined.

Ore does not have to be exhausted at a mine for its reserves to run out. Reserves may be lost if the price of the metal drops to where profitable mining is no longer possible. This is the situation for some Arizona and New Mexico copper mines, and for the Butte mining district of Montana. The

The abandoned Anselmo mineyard in the Butte, Montana, mining district. This underground mining operation was no longer profitable after the 1970s and was therefore closed down.

The shaft to a small copper mine in the Picuris Mountains, New Mexico.

This small mine in the Picuris Mountains, New Mexico, was abandoned after it was no longer profitable to operate.

metal remains abundant in the rock, but mining companies cannot continue to extract it and remain in business. Abandoned mines are no longer reserves and become resources instead.

Environmental concerns can also determine whether a deposit is a reserve or a resource. Some of the issues are whether or not mining should be permitted on public lands, such as national parks, or whether a mining operation is polluting land, air, or water.

In northern Minnesota in the 1970s, the Reserve Mining Company violated environmental standards for water pollution. The company had been hauling taconite, a rock rich in

Drilling to sample the rock surrounding a mining operation provides information on the amount of reserves.

iron, from its mine at Babbitt to its mill 50 miles (80 km) away at Silver Lake. The milling process produced 67,000 tons of waste rock each day, and Reserve Mining disposed of it by dumping it into Lake Superior. When environmentalists realized that this was destroying fish habitats and polluting the region's pure air and water with dust, they asked for a court to order a stop to Reserve Mining's operations. Not pleased about shutting down or having to pay for transporting waste rock to a suitable land site, the company agreed to clean up its operations. Although the mine was not ordered shut, the implications are that certain violations of law or government policy could lead to the closing of a mine and loss of a metal source.

Operating a large mine requires special care for preserving the environment.

Reserves of all metals are limited in number and size. This makes conservation practices necessary so that metals may serve society for centuries to come. Growing world population makes conservation even more important. There are about 4 billion people on earth today, but the year 2000 will see about 7 billion. Each member of this huge population will require a share of world metal reserves.

A first step in protecting metal deposits, then, is population control. This method will be most effective in countries that are the greatest users of minerals, such as the United States and those in western Europe. In the United States, for example, each person consumed about 15 pounds of zinc a year during the 1970s, while each person in Africa used less than half a pound of zinc annually.

In coming years the burden of controlling population will shift to some Asian and Central and South American countries. Nations like India, Mexico, and Brazil are not only growing in numbers of people, but their use of metals in construction, transportation, and electronics is increasing as they try to catch up with the technological development of the United States and western Europe.

Scrap yards have long helped preserve metal deposits. Certain parts of wrecked cars can be reused, thereby reducing the need for new ore to supply axles, gears, and window handles for existing cars that need repairs. The frames of junk cars can be smashed and melted so that the metal can be reused in other products. In fact, scrap metal is almost always added to blast furnaces as an ingredient in steel manufacturing.

Individual consumers have the greatest amount of metal to recycle. One popular conservation method is saving aluminum beverage cans for remelting. Besides conserving aluminum, recycling preserves air quality and energy reserves. The

Using scrap metal to manufacture new products preserves the earth's mineral resources.

Recycling aluminum beverage cans is one way individuals can help make the earth's ore deposits last longer.

Space exploration will help locate new ore deposits on earth and on other planetary bodies.

process of recycling aluminum does not pollute as much or use as much energy as that for extracting aluminum from bauxite.

Even bigger metal resources lie in the 14,000 square miles (36,000 square km) of refuse dumps in the United States. On average, 6 to 9 percent of refuse consists of containers made of iron, aluminum, copper, lead, and zinc. Little, however, is being done about using these consumer scraps because costs of recovering the metals are too high.

Technological advances offer the greatest promise of metal deposits lasting into future centuries. Improved geophysical and satellite exploration can find new reserves. Scientific studies may find substitutes for metals now in demand, or show that a metal not currently mined can eventually have an application in society. Research will enable converting some resources into reserves, including refuse dumps, through economical extraction processes. For the moment, awareness of the earth's limited supply of metals may be the best ammunition in the battle to keep metals from running out. That knowledge forms the foundation for conservation.

Glossary

alloy—any mixture of metals; for example, copper, tin, and zinc melted and mixed form bronze.

crushing—part of the process of separating metal from ordinary rock, in which the metal and rock are crushed to small fragments.

crust—the outermost layer, or shell, of the earth; the top portion of the lithosphere.

deposit—see mineral deposit.

drift—a nonvertical tunnel in an underground mine.

flotation—a process of floating ore powder on a liquid to enable concentrating it and recovering it by skimming.

grade—the quantity, or percentage, of ore-mineral content in rock.

hydrothermal—heated water moving through rock cracks and pores: The heated water is called hydrothermal fluids or solutions.

hypogene—a mineral deposit formed from hydrothermal solutions.

lava—extruded molten rock material flowing from a volcano.

magma—molten rock material formed within the earth; igneous rock forms from the cooling and hardening of magma.

magmatic—derived from magma.

magmatic segregation deposit—an ore deposit formed by the separation and accumulation of minerals crystallized from a cooling magma.

meteorite—stony or metallic material that has fallen to earth from space.

milling—part of the ore process of separating metal from ordinary rock, where the rock containing metal is crushed, or milled, to a fine powder, or "flour."

mineral—a naturally forming inorganic chemical element or compound having a definite chemical composition and physical (atomic) structure.

mineral deposit—a naturally occurring accumulation of metal-bearing mineral(s), or ore minerals, in the earth's crust; some mineral deposits are nonmetallic (not described in this book).

native metal—any element of a metal, such as copper, gold, or silver, found in nature uncombined with other elements.

obduction—the geologic process of thrusting one lithospheric plate above another.

open pit—a surface mine from which rock and ore is continually removed, leaving a large pit in the ground.

ore—naturally occurring material from which a mineral, or minerals, of economic value can be extracted; the mineral(s) that is extracted is also called ore.

ore deposit—a naturally occurring accumulation of minerals that can be extracted for economic purposes.

oxidation zone—an area of a mineral deposit modified by groundwater, whereby oxygen is added to ore minerals to convert them to other types of minerals.

oxidized enrichment—minerals converted to ore minerals by modification in an oxidized zone.

percolate—movement of water through small cracks and pores in rock, usually downward, under the force of gravity.

placer—a mineral deposit formed at the surface by mechanical concentration of mineral particles weathered from rocks; for example, stream water may carry gold grains and drop many of them on a particular part of the stream bed.

plate—segment of lithosphere; plates slowly move a few inches each year.

plate tectonics—the study of the various movements of and interaction among the earth's lithospheric plates.

pluton—a large body of rock formed by subsurface cooling of magma.

reconnaissance—a general exploratory examination, or survey, of the main geologic features of a region.

remote sensing—obtaining geologic information about the earth by distant instruments, such as satellite cameras or radar.

reserves—ore deposits that can presently be mined at a profit.

resources—metal deposits in the earth in more than normal amounts, which are not necessarily profitable to mine at present.

secondary enrichment—a process whereby descending groundwater enriches rock in ore-mineral content; see supergene.

shaft—a passage in an underground mine that is vertical or nearly vertical.

smelter—a place where ore powder is melted to extract the pure metal.

smoker—a slang term given to a vent on the ocean floor that continuously issues a stream of hydrothermal solutions into the ocean.

spreading center—a portion of the earth where two lithospheric plates move apart from each other.

stoping—extraction of ore from an underground mine by blasting walls and ceilings of tunnels.

strategic metals—metals that are vital to the security of a nation but must be obtained largely from mines in foreign countries.

subduction zone—a portion of the earth where one lithospheric plate descends beneath another.

supergene—a mineral deposit formed by groundwater solutions moving downward and enriching the rock in ore minerals.

weathering—the natural destructive process that deteriorates rock, largely by exposure to atmosphere and moisture.

Further Reading

Cronan, D.A. *Underwater Minerals*. Orlando, Fl.: Academic Press, 1980.

Dietrich, R.V., and B.J. Skinner. *Rocks and Rock Minerals*. New York: John Wiley and Sons, 1979.

Fodor, R.V. *Chiseling the Earth: How Erosion Shapes the Land*. Hillside, N.J.: Enslow Publishers, 1983.

Fodor, R.V. *Earth in Motion: The Concept of Plate Tectonics*. New York: William Morrow and Co., 1978.

Govett, G.J.S., and M.H. Govett, eds. *World Mineral Supplies: Assessment and Perspective*. New York: Elsevier Science Publishing Co., 1976

Hutchinson, C.S. *Economic Deposits and Their Tectonic Setting*. New York: John Wiley and Sons, 1978.

Jensen, M.L., and A.M. Bateman. *Economic Mineral Deposits*. New York: John Wiley and Sons, 1981.

Park, C.F. *Earthbound: Minerals, Energy, and Man's Future*. San Francisco, Ca.: Freeman, Cooper & Co., 1981

Park, C.V., and R.A. MacDiarmid. *Ore Deposits*. New York: W.H. Freeman & Co., 1970.

Peters, W.C. *Exploration and Mining Geology*. New York: John Wiley and Sons, 1978.

St. John, J. *Noble Metals*. Alexandria, Va.: Time-Life Books, 1984.

Sawkins, F.J. *Metal Deposits in Relation to Plate Tectonics*. New York: Springer-Verlag, 1984.

Stack, B. *Handbook of Mining and Tunneling Machinery*. New York: John Wiley and Sons, 1982.

Weston, R. *Strategic Materials: A World Survey*. Totowa, N.J.: Rowman & Allanheld Publishers, 1984.

Wolfe, J.A. *Mineral Resources: A World Review*. New York: Chapman and Hall, 1984.

Index

remote sensing, 52, 55
rhenium, 71
rhodium, 71
Roentgen, W.C., 14

S

Salton Sea, California, 25
Schlumberger, C., 56
scrap, 84–85
secondary enrichment, 28–29
silver, 7, 11–12, 23–26, 28, 35, 67–70
smelter, 75–76
smokers, *37, 38*
Steep Rock Lake, Ontario, 5–6
stockpile, 71
Stone Age, 11
strategic metals, 6, 71
subduction, *41,* 43–45
Sudbury, Ontario, 46, 47, *48, 49, 77*
supergene, 28–29, 45

T

taconite, 82

tin, 8, 11–12, 23, 71
titanium, 13–14
tri-state region, 45
tungsten, 71

V

vanadium, 71

W

water table, 28–29

X

x-ray, 14, *15*

Z

zinc, 14, 23, 25, 35, 38, 42, 45, 55, 65, 70, 79, 84, 87
zirconium, 13, 71

C